T0132306

THE WEB

BOOK III

Written and Illustrated by

M. Khalilah Muhammad, Ph.D.

© Copyright 2023 M. Khalilah Muhammad, Ph.D..

All rights reserved. No part of this publication may be reproduced, stored in a retrieval system, or transmitted, in any form or by any means, electronic, mechanical, photocopying, recording, or otherwise, without the written prior permission of the author.

Order this book online at www.trafford.com
or email orders@trafford.com

Most Trafford titles are also available at major online book retailers.

Trafford PUBLISHING® www.trafford.com
North America & international
toll-free: 844 688 6899 (USA & Canada)
fax: 812 355 4082

Our mission is to efficiently provide the world's finest, most comprehensive book publishing service, enabling every author to experience success. To find out how to publish your book, your way, and have it available worldwide, visit us online at www.trafford.com

Because of the dynamic nature of the Internet, any web addresses or links contained in this book may have changed since publication and may no longer be valid. The views expressed in this work are solely those of the author and do not necessarily reflect the views of the publisher, and the publisher hereby disclaims any responsibility for them.

Any people depicted in stock imagery provided by Getty Images are models, and such images are being used for illustrative purposes only. Certain stock imagery © Getty Images.

ISBN: 978-1-6987-1366-3 (sc)

ISBN: 978-1-6987-1365-6 (e)

Library of Congress Control Number: 2022922914

Print information available on the last page.

Trafford rev. 12/09/2022

For my Beautiful Children:

Ashanti
Jameel
Nadirah
Najlah
And
Zafir

Without whom I would never have been
Able to slow down enough
To observe the Facets and
Web of
Life.

ABOUT THE AUTHOR

Born to an ex-military and retired Firefighter and an Educator, Dr. M. Khalilah Muhammad was always reading and examining books and developing her writing skills and the skill of open expression. Due to her father's military background, Dr. Muhammad traveled extensively. She loves visiting new places and experiencing new horizons.

As a child, Dr. Muhammad would spend time in the yard playing with worms and examining the gardens her parents established. There were floral gardens and vegetable gardens in her yard. She would assist in the upkeep of the gardens and the yard.

After a hard day's work, Dr. Muhammad, would sit on the porch and marvel at the bounties of life. One of her favorite was watching the flight of the leaves. The ability to soar and sail while going through the cycles of life amazed her. The second favorite was taking time to enjoy the beach, waves, and life at rest and reflection.

Dr. Muhammad is an epidemiologist. She studies trends of disease and loves traveling and exploring the world. She was told once by a childhood friend to make sure to, "take time and smell the roses". In doing so, she wanted to spend time in the environment and marvel at creation! Dr. Muhammad currently resides in Georgia.

What do you do when you pull in the driveway and see this?

A big spider!

He is in his home!

His home is a large web!

What do you do?

You stop and marvel!

Marvel at the beauty!

Marvel at how nice the web looks!

How did the Spider make this web?

What is in the web with the spider?

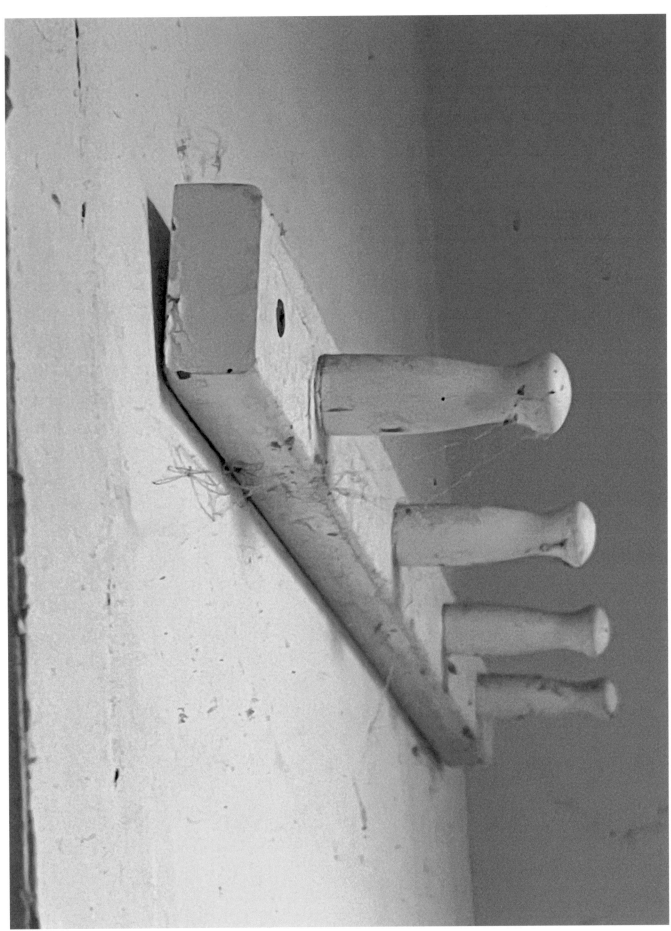

Now we are inside. Look closely.

What type of design is this?

What type of spider lives in this web?

It comes out from the corner of the wall.

A part of it hangs from the ceiling.

It is almost transparent. What does that word mean?

Why is it sticky?

It has other things in it?

The spider eats these wraps. How?

Lightweight insects are the only ones that get caught in the web.

They do not come out of the web.

The web is not disturbed.

I am a human. I am Heavy!

If I walk or run into a spider's web, I will not stick to it.

The spider's web is his home.

If I walk through it, the spider will have to rebuild his home.

How do they start over? Have you run through a web?

What did you do? Write your answers below:

The first string from the spider is long.

It starts from the top part and he looks for his safety.

As he moves from one part of the area to the next part, he

makes sure he will not be eaten.

He spins his web from silk which comes from his tail end.

Can you draw a spider and show where the silk comes from?

How did the Spider make this web?

A spider's web is very frail but sticky. The spider web part that we see is silk. Spiders have glands. This is what makes the silk. The glands are in the abdomen. The silk is sticky but the spider stands where it is not sticky. So they do not get stuck in the web they are making. The spider makes both types of silk. He makes sticky silk and silk that is not sticky.

Can you draw a very thin line? Now draw a spider at the bottom of the line.

What is in the web with the spider?

First the spider lets out a little string of silk. The spider waits until the wind blows and the silk flows with the wind. When the spider's silk sticks to something, the spider will go over to the string and will be adding to it while doing so. The main string is the first string. It is the support string. Sometimes small insects will glide by or get caught in the wind and stick to the web. Look at what happens when it rains on the web!

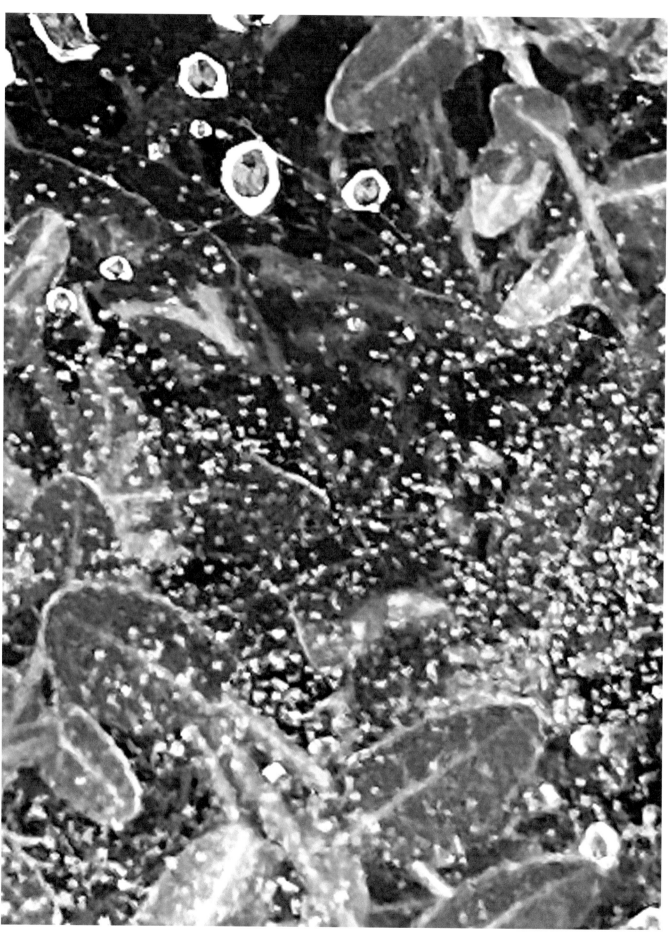

The spider makes seven types of silk that comes from the seven silk glands that they have.

No one spider has all seven glands.

If the spider is a male, he has three types of the glands. His glands make these three types of silk: Dragline, Attachment and Swathing.

Can you draw a picture of a spider? Where are his glands?

If the spider is a female, she produces four types of silk from her four glands. The four types are the same as the male spider but she has another one for the babies! These are the four that she produces: dragline, attachment, swathing and egg sac silk. Look closely in the corner. You will see the egg sacs!

When there is a spider web that is attached to a board or any thing, it is called attaching threads.

Attaching Threads are the threads that attach to a thread or an object.

The mother spider has the egg sac and the silk she makes is the egg sac silk. Draw an egg sac in the first box below. Draw a spider coming out of the sac in the second box below. Draw a baby spider in the third box below. Draw a spider in its own home in the fourth box.

1.	2.
3.	4.

The dragline silk is made by both male and female spiders. Draw a spider web and show where the dragline is located:

All spiders have sticky silk which means that they have an additional set of glands that make sticky silk and sticky droplets for the web. That is what makes it sticky! Many are found in your garage!

Here are some types of Spider Webs!

Find the type of spider web listed on each page. Draw it and put your picture below the type of web:

The Spiral orb Web

The Tangle web. This web is also called the Cob web.

The Dome web. This is also called the Tent web.

The Funnel Web which can be primitive or modern. Can you find a photo of a funnel web spider? Draw a funnel web and put your photo of the spider inside of the web.

The Tubular web is often seen on the bottom of the trees and on the ground. Can you draw a web at the bottom of the tree that you might see when you go outside to play?

Finally, **the Sheet web** which uses the sticky webs in a horizontal way.

Do you know what horizontal is? What is the opposite?

Word Search

FIND THE WORDS BELOW ....

C	L	R	G	F	J	S	P	I	R	A	L	V	T
A	J	S	P	I	D	E	R	O	G	T	Y	H	U
S	T	I	C	K	Y	V	X	B	A	T	E	I	B
G	R	A	B	R	J	F	I	C	E	A	N	X	V
G	A	T	N	A	L	H	W	E	B	C	I	C	L
E	N	E	J	G	T	A	G	I	N	H	L	O	A
A	S	E	A	A	L	S	I	L	K	M	G	B	R
N	P	H	W	M	L	E	U	I	A	E	A	W	A
H	A	S	M	H	U	M	A	N	M	N	R	E	E
F	R	A	I	L	A	D	J	G	A	T	D	B	M
T	E	N	T	W	E	B	N	A	A	N	M	S	O
E	N	G	F	U	N	N	E	L	W	E	B	A	D
S	T	R	I	N	G	P	Y	R	I	F	O	R	M

COBWEB

FUNNELWEB

TANGLE

DOME

SHEET

WEB

CEILING

HUMAN

SILK

GLANDS

DRAGLINE

PYRIFORM

ATTACHMENT

EGGSAC

SPIDER

TRANSPARENT

SPIRAL

STICKY

TUBULAR

FRAIL

STRING

SWATHING

How did your word search go? How many words did you find on your own?

Did your parent(s) help you? How many did they help you find?

How do Spiders help you?

How do Spiders hurt you?

What is the name for the spider in Science?

Do you have pictures of spiders? If so, attach them...

Do you have pictures of spider webs? If so, attach them...

THANKS FOR YOUR SUPPORT!

Printed in the United States
by Baker & Taylor Publisher Services